MATHEMATICIANS are counting the STARS!

Written by
SASKIA GWINN

Illustrated by
ANA ALBERO

MAGIC CAT 🐾 PUBLISHING

"Other magnificent mathematicians of course! Mathematicians do the MOST marvellous things, including..."

Finding out how fast animals are,

learning about earthquakes,

looking inside your body,

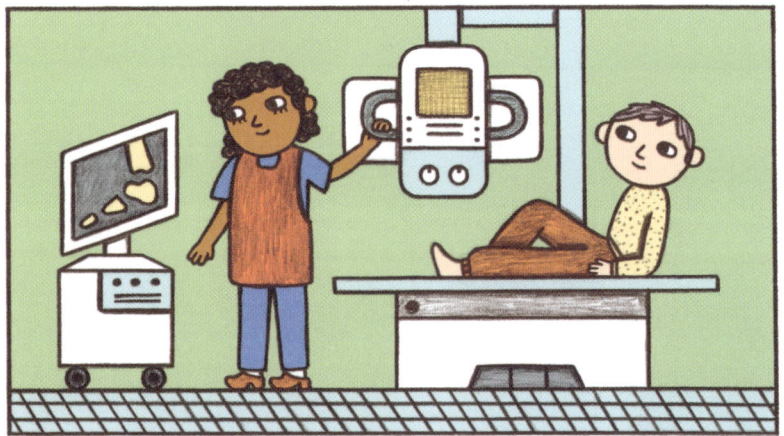

making cars cool,

and EVEN exploring space.

"Woah. To look for aliens?"

"Yep. Just wait and see..."

MATHEMATICIANS ARE FINDING OUT HOW FAST ANIMALS ARE

Mathematicians work out how SPEEDY animals are with their FANTASTIC FORMULAS.

GALILEO GALILEI figured out how to measure speed!

So, now we can calculate how quickly animals run!

But amazing DR MYRIAM HIRT discovered you have to go even further to work out an animal's TOP SPEED.

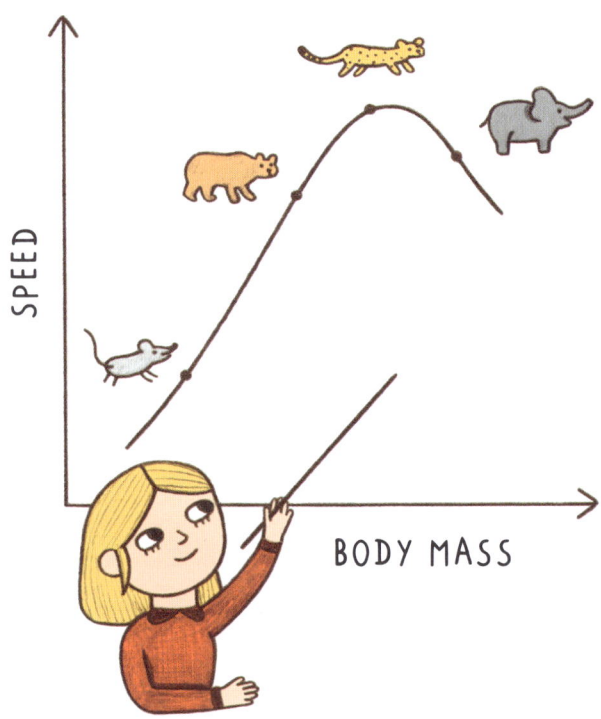

Myriam used a PHENOMENAL maths formula to figure out that to be the fastest runner, you cannot be...

Too BIG

or too SMALL...

But rather, somewhere in between.

ZOOOOM!

Large animals run out of energy before they can gather top speed, while small ones can't cover enough distance with their little legs!

Myriam's mega maths model can even be used to find out how fast dinosaurs ran!

"That is cool. What else do mathematicians do?"

"Well, they look inside our bodies..."

MATHEMATICIANS ARE FINDING OUT WHAT'S INSIDE YOUR BODY

Mathematicians use MIND-FIZZING maths to help discover what makes you... you!

Dr C. R. RAO learned about numbers as a little boy. By the age of six, he knew his 20 times table!

While a student at Cambridge University, Dr Rao used MESMERIZING maths to map chromosomes in mice. Understanding chromosomes helps us to know which bits of our DNA come from which ancestor.

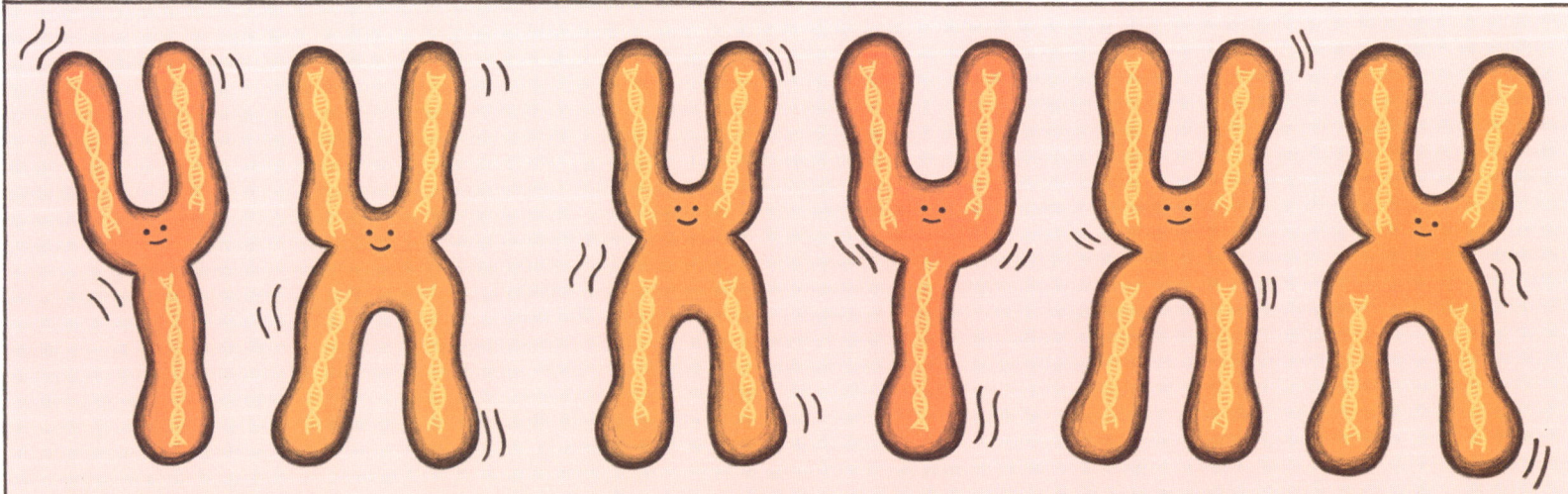

Chromosomes live inside our body's cells. They contain DNA. DNA tells our cells what to do, and is unique to each and every one of us.

It was ROSALIND FRANKLIN whose earlier work allowed us to understand the structure of DNA. She used an X-ray photo and some world-shaking, life-changing maths...

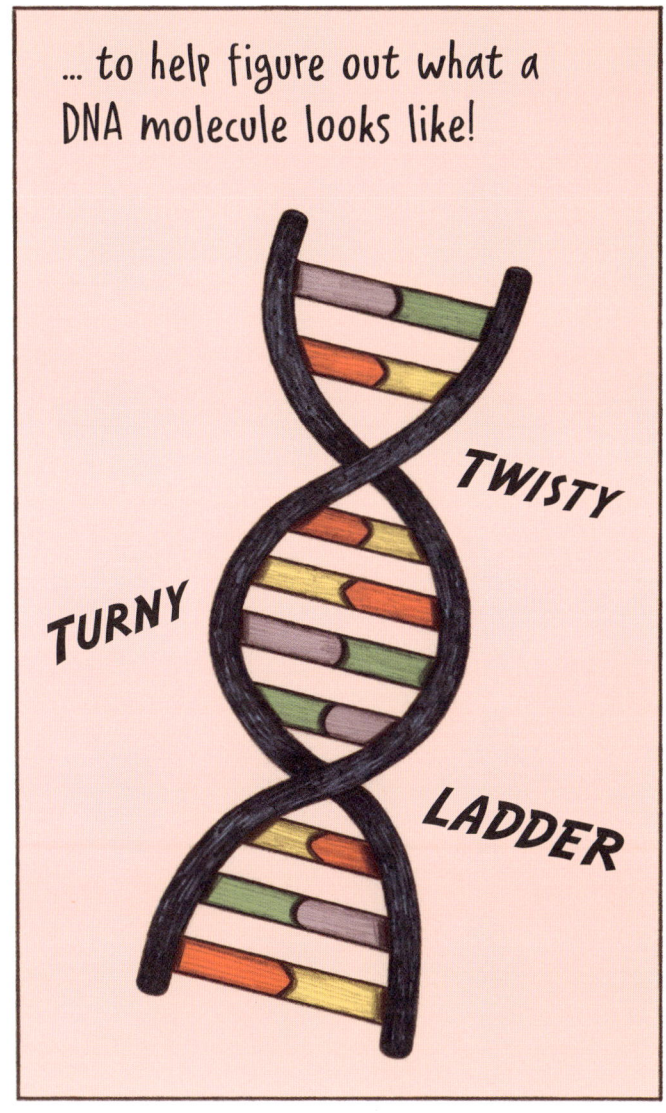
... to help figure out what a DNA molecule looks like!

TURNY TWISTY LADDER

Knowing about DNA inside our bodies helps mathematicians — and scientists — find cures for diseases.

Can mathematicians see inside anything else?!

Yep. Earth!

MATHEMATICIANS ARE LEARNING ABOUT EARTHQUAKES

Mathematicians use their MAGNIFICENT measuring skills to find out IMPORTANT things about earthquakes. They discover...

Where earthquakes start — EPICENTRE, HYPOCENTRE, FAULT, HERE!

and how fast earthquakes travel. SEISMIC WAVE

Seismologist INGE LEHMANN used maths to find out something AWESOME about earthquakes.

BEND AND BOUNCE!

Earthquakes release waves of energy called SEISMIC WAVES and... some of the waves are bent!

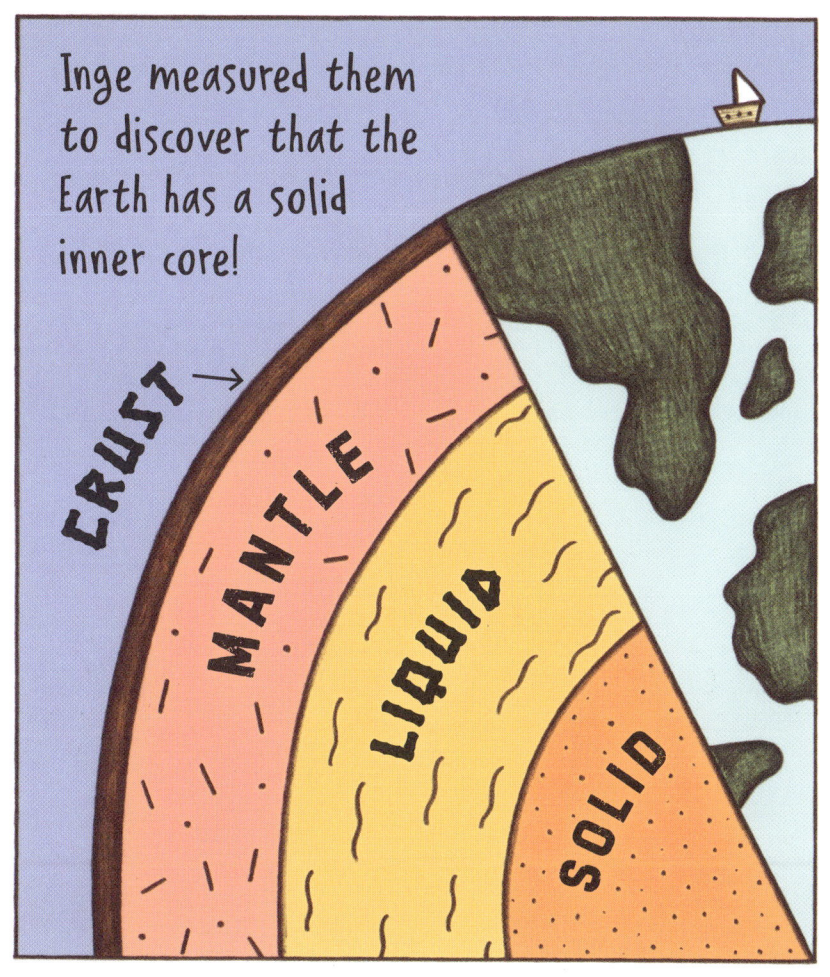
Inge measured them to discover that the Earth has a solid inner core!
CRUST MANTLE LIQUID SOLID

ANDRIJA MOHOROVIČIĆ used his super maths skills to find out the speeds at which earthquake waves travel.

Discovering facts about earthquakes can help us to control some of the HORRIBLE damage they cause.

What else do mathematicians know about?

Cars.

MATHEMATICIANS ARE MAKING CARS COOL

Mathematicians use their EXTRAORDINARY mathematical minds to give cars...

Super-cool shapes,

talking maps, "STRAIGHT ON"

and the use of ELECTRICITY.

Electric cars burn fewer gases that are harmful to our planet.

RALPH BRAUN worked hard to make wheelchair lifts for cars, buses and vans!

GLADYS WEST'S work with maths helped to invent GPS, which our vehicles and smartphones use to tell us which way to go.

"IN 450 METRES, TURN LEFT"

Today, mathematicians are EVEN inventing cars that can drive themselves!

Do mathematicians make buildings too?

They do...

MATHEMATICIANS ARE CONSTRUCTING SKYSCRAPERS

Engineers and architects use MARVELLOUS maths to help create the TALLEST and most beautiful buildings in the world.

They know about SHAPES,

CURVES,

PATTERNS...

and HEIGHT.

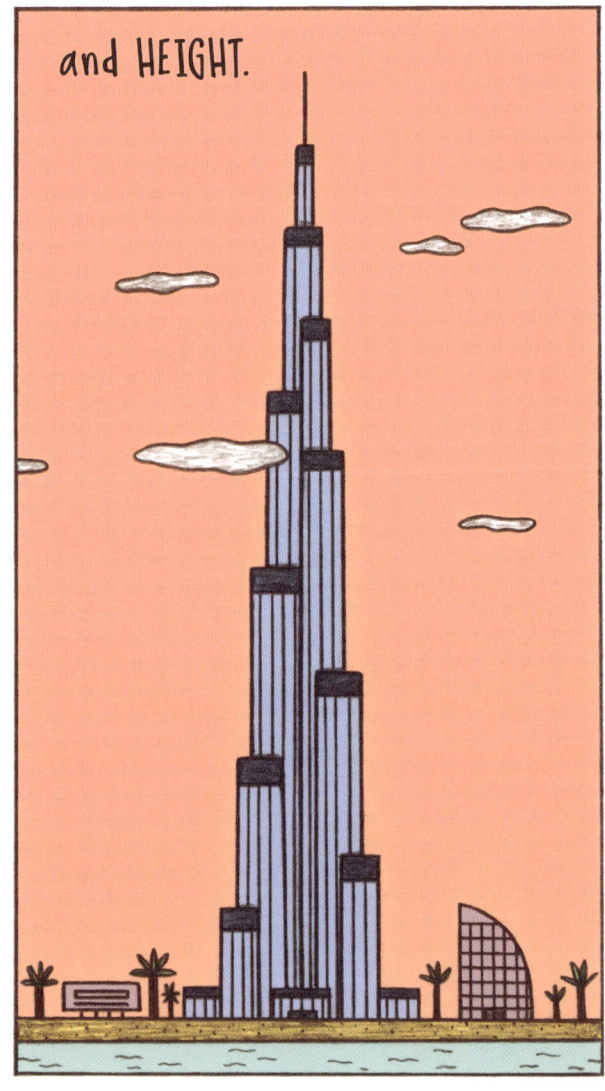

ROMA AGRAWAL uses maths to figure out how to stop skyscrapers from swaying in the wind too much.

She worked out how to construct the very top of the tallest building in Britain...

the SHARD.

KAZUYO SEJIMA is an architect who uses maths to give buildings special shapes such as SQUARES and CUBES.

It's not just ENGINEERS and ARCHITECTS who make brilliant buildings by using maths. CARPENTERS, CONSTRUCTION WORKERS and BUILDERS all use maths too!

What other amazing things do mathematicians make?

Computers and games!

MATHEMATICIANS ARE CODING COMPUTER GAMES

Mathematicians use MIND-BLOWING maths to code INCREDIBLE computer games.

ADA LOVELACE was interested in how things work. She reviewed plans of THIS machine and figured out...

ANALYTICAL ENGINE

...that it could use numbers to follow instructions!

Many years later, this led to the invention of...

the computers you play games on!

JERRY LAWSON loved video games SO much, he taught himself to create new ones!

Years ago, video-game consoles had built-in games. Jerry led a team of people using cool maths calculations to develop the first consoles which played games stored on cartridges.

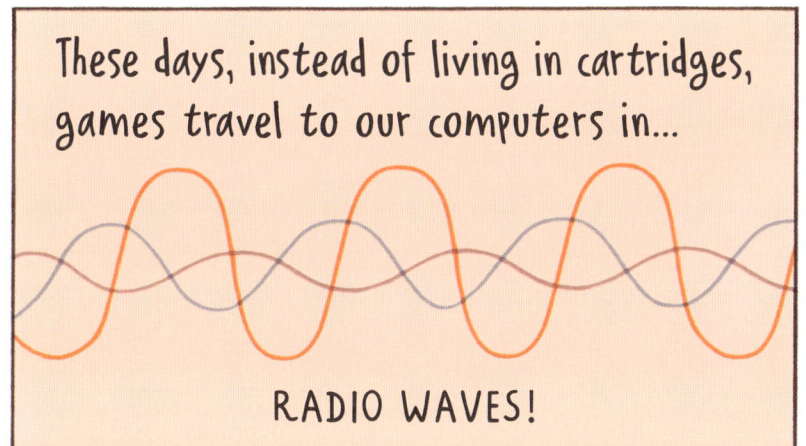

These days, instead of living in cartridges, games travel to our computers in...

RADIO WAVES!

through super-duper fast fibre-optic CABLES buried under the sea!

All the way to your homes!

Mathematicians helped make this happen!

"Computer maths is smart!"

"Yes! And it's getting smarter..."

MATHEMATICIANS ARE INVENTING ARTIFICIAL LIFE

Computer scientists use clever maths to make ASTOUNDING algorithms. Algorithms are sets of instructions that tell computers what to do.

They are used to invent machines that keep getting...

cleverer
SPARKLY CLEAN!

and cleverer

and cleverer!

Mathematicians wonder if computers can do things that the human brain can. We call this study Artificial Intelligence or AI.

HELLO!

JOHN MCCARTHY invented the term AI. He programmed computers to play chess.

RANA EL KALIOUBY is programming computers to recognize how we are feeling!

Computers cannot think for themselves like we do, but because of clever maths they can beat world-champion HUMANS at chess.

"I WIN!"

But brilliant human brains will always come up with colossally cool ideas, too, that computers can't come up with!

"So, mathematicians teach computers stuff... WHO or WHAT teaches mathematicians?"

"NATURE!"

MATHEMATICIANS ARE NOTICING NATURE PATTERNS

Mathematicians use wonderfully WILD maths to find patterns in NATURE.

BENOIT MANDELBROT used maths to discover that patterns can go on FOREVER. Patterns that repeat each other over and over and OVER again are called FRACTALS.

$$D = \frac{\log N}{\log(\frac{1}{r})}$$

You can see them on... PINE CONES,

PLANTS,

PINEAPPLES

and PEACOCKS!

ALAN TURING used his WHIZZY MATHS MIND to figure out why...
a giant pufferfish has its funky PATTERNS,

a shell has its SWIRLS,

and a leopard has its SPOTS!
The spots are created by chemical reactions inside their bodies.

Now mathematicians know there are repeated patterns inside YOU! This helps doctors learn how to make people better.

What else can mathematicians discover?
How to make the fastest things in the sky!

MATHEMATICIANS ARE MAKING MEGA-FAST PLANES

Tremendously TERRIFIC mathematicians are helping to build the fastest things in the sky, such as...

HELICOPTERS, AEROPLANES AND MEGA-FAST MACHINES!

ZIP ZAP ZOOOOM!

Like the tremendous inventors WILBUR and ORVILLE WRIGHT.

ORVILLE WILBUR

They designed, built and FLEW the first piloted and powered airplane.

THEODORE VON KÁRMÁN was curious about planes. He used maths to work out what shape the wings needed to be to go SUPER FAST.

Supersonic planes could WHOOSH through the sky faster than the speed of sound!
750 MPH
THUNDER THUNDER BOOM!
But supersonic planes make a supersonically LOUD sound too!

So, mathematicians like CHRISTINE DARDEN worked out how to make them quieter!

We can't fly in supersonic planes over land (they are still TOO NOISY!).

But, mathematicians are figuring out if it will be possible.

Does that mean mathematicians predict... the FUTURE?
Pretty much!

MATHEMATICIANS ARE PREDICTING THE FUTURE

Mathematicians use SUPER-HERO maths to predict the FUTURE. They forecast what will happen to our planet if we don't look after it.

SYUKURO MANABE used PLANET-SPINNING maths to warn us that the world will get hotter.

He told us we are creating too much carbon dioxide.

CHRISTIANE ROUSSEAU brought mathematicians TOGETHER to discuss many things, including: what will happen if the world gets TOO HOT?

⇶ MATHEMATICIANS ARE COUNTING THE STARS ⇶

Mathematicians count things with their spectacular SPACE maths.

KATHERINE JOHNSON couldn't get enough of counting and calculating.

She used some COSMIC maths to figure out flight paths to the Moon.

Mathematicians count stars too. They think there are over 100 billion stars in a galaxy called the Milky Way.

1 2 3 4 5 6 7 8 9 10 11 12

If each star has lots of planets around it like our own star (the Sun) does...

then there are hundreds of billions of planets out there that we haven't yet seen!

But stars are LIGHT-YEARS away.

HENRIETTA SWAN LEAVITT helped us to work out how far away they are.

A star's light is so BRIGHT that it's hard to see what planets might be nearby!

ZING!

So mathematicians helped invent a machine that blocks out their BLAZING light.

This means we can see the planets that are near them! And wonder whether any are home to alien LIFE...

"So, MATHEMATICIANS are looking for aliens?!"

"Yes!"

MATHEMATICIANS ARE LOOKING FOR ALIENS

Now, we know there are probably hundreds of BILLIONS of planets out there, so mathematicians are looking for signs of life in SPACE.

But – we are a LONG way from finding space ALIENS just yet...

First, we have to find planets outside our SOLAR SYSTEM. They are called EXOPLANETS.

Mathematicians look for them with their incredible maths models!

SARA SEAGER uses maths to find planets that are a LONG WAY AWAY.

And MICHELLE KUNIMOTO measures COSMIC LIGHT CURVES to find new planets.

CHECK OUT these COOL-LOOKING exoplanets. Mathematicians helped to find them too!

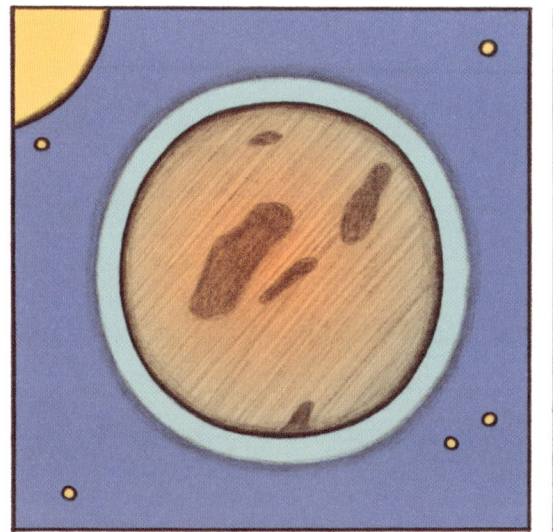

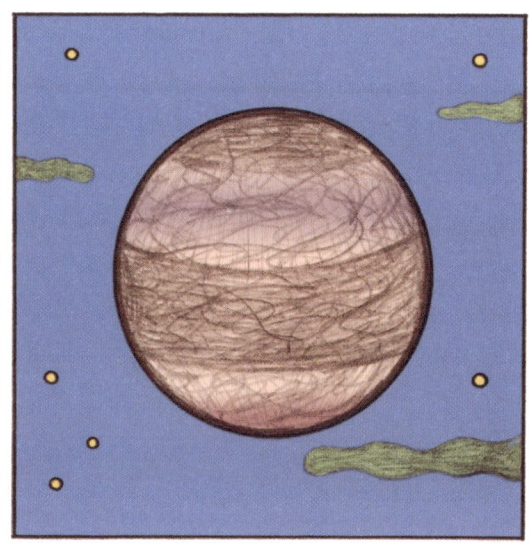

Maybe one day, mathematicians WILL figure out where aliens live too...

FINALLY!

Mathematicians like ME?

It could be you! Because...

MATHEMATICIANS ARE LIKE YOU

Mathematicians are always looking at NUMBERS, SHAPES, PICTURES and MEASUREMENTS to create magical maths. And they began by...

peering at patterns,

building with bricks,

asking about animals,

HOW FAST?!

reading about rockets...

WHICH MATHEMATICIAN INSPIRES YOU?

ASTRONOMER

GALILEO GALILEI
Born 1564 · Died 1642

Galileo was a mathematician, astronomer and natural philosopher who figured out how to measure how fast things travel. He also explored space with his terrific telescope and discovered moons orbiting Jupiter. At first, he thought the moons were stars.

STATISTICIAN

DR C. R. RAO
Born 1920

Dr Rao is an exceptional statistician whose amazing maths models have inspired people all over the world. Some super-famous maths formulas are named after him. He was awarded the 2023 Prize in Statistics for his outstanding work with maths.

SEISMOLOGIST

INGE LEHMANN
Born 1888 · Died 1993

Inge was a seismologist who used maths to discover the inner core of the Earth. She realized that earthquakes release energy waves and measured them, too, noticing that some waves were bent. This is because Earth has a solid inner core and a liquid outer core!

RESEARCHER

DR MYRIAM HIRT
Born 1988

Dr Myriam used a fantastic formula to find out an animal's top speed. She worked with a team to discover that the biggest beasts, such as elephants and even the terrifying T-Rex, ran out of energy before they reached their fastest run. Huge muscles don't always help you to run the fastest!

CHEMIST

ROSALIND FRANKLIN
Born 1920 · Died 1958

Rosalind's amazing maths contributed to working out the structure of DNA. She collected data from X-ray photos. The peculiar patterns in an X-ray called Photograph 51 helped to make the magnificent discovery. Rosalind was not credited for her work until years after she died.

SEISMOLOGIST

ANDRIJA MOHOROVIČIĆ
Born 1857 · Died 1936

Andrija found out epic things about earthquake waves, such as how fast they travel. One of his main discoveries was that they move faster as they get closer to the centre of the Earth. He investigated further to find out what earthquake waves tell us about planet Earth.

ENGINEER

RALPH BRAUN
Born 1940 · Died 2013

Ralph was diagnosed with a muscle disability at just six years old and began using a wheelchair in his teens. As a kid he built cool stuff in his dad's garage and later developed ground-breaking wheelchair access in cars and other vehicles. He built a motorized scooter too!

ENGINEER

ROMA AGRAWAL
Born 1983

Roma is a structural engineer who used amazing maths to help design the very top of the UK's tallest skyscraper, the Shard. She is a storyteller too and writes books that tell us about the history of construction and how inventions have changed the world.

MATHEMATICIAN

ADA LOVELACE
Born 1815 · Died 1852

Ada worked with another awesome mathematician called Charles Babbage. He designed a machine which Ada realised could be programmed to follow instructions. The notes she wrote about it are regarded as the world's first computer program.

MATHEMATICIAN

GLADYS WEST
Born 1930

Gladys is a magnificent mathematician who helped invent the technology that contributed to the invention of GPS (Global Positioning System). She looked at information collected by satellites and created clever calculations that were used to make computer models of the Earth's surface.

ARCHITECT

KAZUYO SEJIMA
Born 1956

Kazuyo is an award-winning architect who also works with Ryue Nishizawa to make beautiful buildings. They often use shapes such as squares and cubes in their dazzling mathematical designs that have created houses, offices, art centres and education sites.

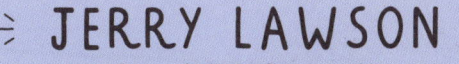

ENGINEER

JERRY LAWSON
Born 1940 · Died 2011

In his youth, Jerry loved to tinker with electronic things and would repair TV sets during his high-school days. He later became an engineer, and in the 1970s he invented a video-game console that played different games using special cartridges.

COMPUTER SCIENTIST

JOHN MCCARTHY
Born 1927 · Died 2011

John McCarthy is said to be the person who came up with the words 'Artificial Intelligence', shortened to 'AI'. He thought about how computers compare to human brains and used clever maths to make computers cleverer too. He even programmed one to play chess against another computer.

MATHEMATICIAN

BENOIT MANDELBROT
Born 1924 · Died 2010

Benoit Mandelbrot discovered important things about patterns. He has a maths concept named after him called the Mandelbrot set, which shows how a pattern can be repeated over and over again. He received many awards and honours for his amazing mathematical work!

AEROSPACE ENGINEER

THEODORE VON KÁRMÁN
Born 1881 · Died 1963

Theodore was a mathematician who was super interested in aircraft. He studied the impact of airflow on flying machines, which aided in the invention of supersonic flight. He also worked on wind tunnels that would be used for testing airplane models.

COMPUTER SCIENTIST

RANA EL KALIOUBY
Born 1978

Rana looks at how computers can detect human emotions so that they can tell how we are feeling. One day, computers might even be able to work out if we have an illness, which would help to save human lives.

MATHEMATICIAN

ALAN TURING
Born 1912 · Died 1954

Alan was a world-changing mathematician. He worked out computer maths, biological maths and even secret codes! He also used his remarkable maths skills to figure out why animals have patterns such as spots, stripes and other interesting markings.

AEROSPACE ENGINEER

CHRISTINE DARDEN
Born 1942

Christine loves finding out how things work. She has spent many years studying something called the 'sonic boom'. It is the mega-loud supersonically swift planes make when they fly faster than the speed of sound. Christine was an aerospace engineer at NASA.

CLIMATOLOGISTS

SYUKURO MANABE
Born 1931

Syukuro began his career by focusing on weather forecasting. He used maths and computers to create the first model of the Earth's climate, and worked out that the Earth would get hotter. He later won a Nobel Prize for his global warming predictions.

MATHEMATICIAN

KATHERINE JOHNSON
Born 1918 · Died 2020

Katherine always loved using numbers. While working for NASA, she used some cool calculations and jaw-dropping geometry (the study of points, lines and solids) to work out how to help astronauts zoom safely to the Moon.

ASTROPHYSICIST

SARA SEAGER
Born 1971

Sara looks for signs of life on planets outside our solar system. She works on many projects, including one called 'Starshade', which is a specially shaped screen that blocks a star's light to help scientists to see any planets orbiting around the star.

MATHEMATICIAN

CHRISTIANE ROUSSEAU
Born 1954

Christine helped organized a global project called Mathematics of Planet Earth. It encouraged mathematicians around the world to discuss important issues about Earth, such as climate change, earthquakes and even pandemics (when a disease spreads around the world).

ASTRONOMER

HENRIETTA SWAN LEAVITT
Born 1868 · Died 1921

Henrietta was an astronomer who studied stars. Her research into the brightness of stars led others to be able to calculate the distance of stars from Earth and also to estimate the size of the Milky Way. She discovered four novas (exploding stars) and around 2,400 'variable' stars herself.

ASTRONOMER

MICHELLE KUNIMOTO
Born 1993

Michelle is looking for life on other planets. She uses mathematical measurements that show how bright stars are to help her find new planets that orbit them. She has already discovered thousands of new planets in space! She became interested in astronomy as a child.

FURTHER READING

The Girl With a Mind for Math
By Julia Finley Mosca and Daniel Rieley

Hidden Figures
By Margot Lee Shetterly

Little People, Big Dreams: Ada Lovelace
By Maria Isabel Sanchez Vagara

Fantastically Great Women Who Changed the World
By Kate Pankhurst

Fantastically Great Women Who Saved the Planet
By Kate Pankhurst

Finding the Speed of Light: The 1676 Discovery That Dazzled the World
By Mark Weston

Counting on Katherine
By Helaine Becker

Nothing stopped Sophie
By Cheryl Bardoe

The Boy Who Loved Math
By Deborah Heiligman

Maryam's Magic: The Story of Mathematician Maryam Mirzakhani
By Megan Reid

The Questioneers: Rosie Revere, Engineer
By Andrea Beaty

Math Curse
By Jon Scieszka

For Etta ~ S.G.

This book is dedicated to all those people out there who, like me, struggled to learn maths as a kid. ~ A.A.

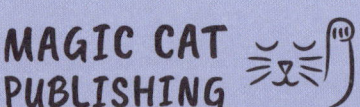

Mathematicians are Counting the Stars © 2024 Lucky Cat Publishing Ltd
Written by Saskia Gwinn
Text © 2024 Magic Cat Publishing
Illustrations © 2024 Ana Albero
First Published in 2024 by Magic Cat Publishing, an imprint of Lucky Cat Publishing Ltd,
Unit 2 Empress Works, 24 Grove Passage, London E2 9FQ, UK

The right of Ana Albero to be identified as the illustrator and Saskia Gwinn to be identified as the author of this work has been asserted by them in accordance with the Copyright, Designs and Patents Act, 1988 (UK).

No part of this publication may be reproduced, stored in a retrieval system, or transmitted, in any form, or by any means, electrical, mechanical, photocopying, recording or otherwise without the prior written permission of the publisher or a licence permitting restricted copying.

A catalogue record for this book is available from the British Library.

ISBN 978-1-915569-24-0

The illustrations were created in pencil and coloured digitally
Set in Ana Albero font, Delivery Note and Lilly

Published by Rachel Williams and Jenny Broom
Designed by Maisy Ruffels
Edited by Rachel Williams

Manufactured in China

9 8 7 6 5 4 3 2 1